Fernando Sanford

Some Observations upon the Conductivity of a Copper Wire

in Various Dielectrics

Fernando Sanford

Some Observations upon the Conductivity of a Copper Wire in Various Dielectrics

ISBN/EAN: 9783337418274

Printed in Europe, USA, Canada, Australia, Japan

Cover: Foto ©berggeist007 / pixelio.de

More available books at **www.hansebooks.com**

LELAND STANFORD JUNIOR UNIVERSITY PUBLICATIONS

·

———

STUDIES IN ELECTRICITY

No. 1

———

SOME OBSERVATIONS

UPON THE

CONDUCTIVITY OF A COPPER WIRE

IN VARIOUS DIELECTRICS

BY

FERNANDO SANFORD, M. S.

Professor of Physics in the Leland Stanford Junior University

PALO ALTO, CALIFORNIA
PUBLISHED BY THE UNIVERSITY
SEPTEMBER, 1892

SOME OBSERVATIONS

UPON THE

CONDUCTIVITY OF A COPPER WIRE

IN VARIOUS DIELECTRICS

It is probable that no known electrical phenomenon offers greater difficulty of explanation at the present time than the phenomenon of metallic conductivity. Faraday has given reasons for believing that it is essentially related to the phenomenon of static induction, and Maxwell's theory would make both phenomena alike dependent upon an elastic fluid, or ether, which permeates both the conductor and the dielectric around it. According to this view, an electric charge consists of a displacement of this elastic fluid, which, on account of its elasticity, tends to return to its original position, while a current consists of a displacement of this same fluid throughout the entire length of a closed conducting circuit, in which case the tendency to return to its original position does not appear. It would, accordingly, appear that this electric elasticity of the ether disappears in conductors, or at the bounding surface between conductors and dielectrics. Since apparently the same inducing force may induce different charges in different dielectrics, it seems necessary to assume that the elasticity of the ether varies in different dielectrics. Since, also, the ether displacement for a given force will be inversely as its elasticity, it follows that the

3

specific inductive capacity of a dielectric will be proportional to the reciprocal of the elasticity of the ether in
this dielectric when subjected to electric stress.

It is now well known that light waves are caused by
electric stresses in this same ether, and it seems necessary
to assume that they, too, depend upon this same electric
elasticity, and, accordingly, that a fixed relation should
exist between the velocity of light in a dielectric and the
specific inductive capacity of the dielectric. This relation,
as determined by Maxwell and Helmholtz, would make
the optical index of refraction proportional to the square
root of the dielectric constant. While it is true that this
relation has not been extensively verified by experiment,
it is also true that different determinations of the
dielectric constant have given very different results for
the same substance.

So far as I am aware, it has been assumed that the
electric conductivity of a wire or other metallic conductor
is uninfluenced by the nature of the dielectric in its field
of force; and depends only upon the nature and temperature of the conductor. The close relation which seems
to exist between the current in a conductor and the phenomenon of induction in a dielectric, and the apparent
fact that the ether displacement in a conductor carrying
a current is caused by a lateral stress communicated to it
by the ether in the dielectric surrounding the conductor,
has, however, made it seem to me probable that the amount
of this displacement for a given force might also be modified by the nature of the surrounding dielectric, or that
the conductivity of a given wire might vary in different
dielectrics.

More than a year ago, I made some experiments to see
if this phenomenon of variable conductivity could be

observed, but with only negative results. Wishing to test the matter more fully,I had made a piece of apparatus by means of which I have been able to observe the suspected phenomenon in liquid and gaseous dielectrics, and by means of which I hope to accumulate much more valuable data concerning the action of dielectrics upon conducting wires.

The apparatus used consists of a cylindrical copper tube 120 cm. long and 2.5 cm. in internal diameter. The ends of the tube are closed air tight by copper plates, which are provided with stopcocks for filling and emptying the tube. To the inside of one end plate is fastened a copper wire 1 mm. in diameter, which passes length wise through the center of the tube and out through an insulated opening in the other end plate, where it is soldered to a piece of lamp cord, made of many small wires, by which it is connected with the Wheatstone's Bridge. Another similar piece of lamp cord is soldered to the end of the tube through which the wire passes, and is likewise connected with the bridge. Midway between the ends of the tube is another tube 5 cm. long and 1 cm. in diameter entering it from the side. This tube serves to admit the thermometer by which the temperature of the interior of the large tube is measured.

The current used for making the measurements passes one way through the tube and returns through the wire. By this arrangement, the entire field of force of the current to be measured is confined within the tube, and the whole of the dielectric concerned in its transmission can be changed at will.

The measurements were made with a Hartmann & Braun combined resistance box and bridge, with bridge arms of 1:1000. The smallest resistance used was .1 ohm, which with the above combination, represented a resistance in the wire and tube of .0001 ohm. The galvanometer used is the physiological galvanometer of DuBois Reymond, manufactured by the Geneva Society Construction Company. It is provided with a concave mirror and ground glass scale, and the usual method of observing the deflection of a spot of light reflected from the mirror upon this scale was used. This deflection was plainly noticeable for a change of resistance corresponding to .0001 ohm in the wire and tube, and by reversing the current several times, it was possible to obtain a deflection for a change corresponding to one-fifth of the above. The measurements were accordingly estimated with a fair degree of accuracy to .00001 ohm. As the combined resistance of the wire and tube was about .0335 ohm, the average of any set of measurements was certainly not .03 of one per cent wrong. The zero method of measurement was adopted throughout. The current used was obtained from a battery of 32 silver chloride cells, usually through 100 ohms resistance between the battery and the bridge, and the strength of current through the wire and tube was always between the extremes of five and eight milli-amperes. The measurements were all made at the temperature of the room, and as this changed very slowly, only a few measurements could be made in a day.

COMPARISON OF SOME LIQUID DIELECTRICS WITH AIR.

The first set of measurements was made with air and wood alcohol as dielectrics. I thought that, on account of the high specific inductive capacity attributed to this liquid, it would be as likely as any to give results differing from air. In this I was disappointed. After drawing a curve representing the resistance of the wire at different temperatures in air, the corresponding curve for wood alcohol was found to practically coincide with it.

The wood alcohol was then poured out of the tube, but was not carefully drained off, and the tube was filled with petroleum. An increase in the resistance of the wire was at once noticed. Eleven measurements were made with this dielectric during two days, March 4th and 5th, the liquid being poured off six times and measurements made in the air in the meantime, but the same liquid being poured back each time. The curve representing these eleven measurements was found to indicate a resistance .00008 ohm greater than that represented by the curve for the air readings. As the average resistance of the wire and tube at the temperature at which the measurements were made was about .03350 ohm, this difference represented about .27 of one per cent of the whole resistance. It was noticed that the liquid had a cloudy appearance, looking neither like wood alcohol nor petroleum, but as the wood alcohol had caused no variation in the resistance, I supposed this increase of resistance to be due to petroleum.

On the following day, March 6th, the tube having been carefully drained out and dried was filled with pure petroleum and a new set of measurements was begun. These indicated at once a decrease of resistance in the wire. As this seemed to contradict the former measurements, I thought that some of the contact resistances about the apparatus were changing. Measurements were accordingly made at irregular periods from March 6th to April 5th, the battery and galvanometer being detached and used for other purposes and the apparatus being allowed to stand from March 14th to 28th. In all, 34 measurements were made in air and 22 in petroleum, the dielectric in the meantime being changed five times, and the temperature changes being from 13.6° C. to 26° C. The whole series was entirely consistent. Not a single reading made with the wire in either dielectric crossed the curve for the other dielectric. The average difference in the resistance of the wire in the two dielectrics was .00006 ohm, which corresponds to about .18 of one per cent of the whole resistance, the conductivity of the wire being that much greater in petroleum than in air.

The curves made for these dielectrics are shown in Fig. 1.

They were drawn as follows : The points representing the resistance at the different temperatures were platted on cross section paper, none of the points for either dielectric being omitted, and a line was drawn through them with a pencil and ruler as nearly as possible in the true direction of the curve, which did not vary perceptibly from a straight line. The distance of each point from the line was then measured, and the algebraic sum of these distances divided by the number of the points taken was assumed as representing the true distance of the curve above or below the trial curve from which the distances were measured. The final curve was then carefully drawn parallel to the other curve at this distance above or below it, as the conditions required.

The above measurements made it seem probable that the increased resistance with the first petroleum was due to the wood alcohol which was mixed with it. To make this certain, the petroleum was poured off, leaving a little in the tube, and the tube was then filled with wood alcohol. Thirty-one measurements with the wire in this liquid were made on April 5th, 6th, and 7th, the liquid being poured out twice in the meantime and twelve measurements made in the air. The thirty-one successive measurements made for the liquid represented an increase of resistance over the former curve made for air of .00007 ohm, while the twelve air measurements averaged only .000004 ohm from the former curve. This liquid, which consisted principally of wood alcohol and contained only a little petroleum, gave a resistance to the wire .2 of one per cent greater than it had been in air; while the liquid used a month before, consisting principally of petroleum with only a little wood alcohol, gave an increased resistance to the wire of .27 of one per cent.

This peculiar effect of using two dielectrics which did not seem to combine with each other was observed later with wood alcohol and benzine. On May 11th and 12th another series of five measurements was made with wood alcohol which seemed to give the wire an increased resistance of .02 of one per cent, a series of seven measurements with the wire in benzine gave an increased resistance of .06 of one per cent, and a series of nine measurements when the tube contained nearly equal parts of wood alcohol and benzine gave an increased resistance of .15 of one per cent.

The other liquids tested were absolute alcohol, ordinary 90 per cent alcohol, carbon bi-sulphide, distilled water, and a mixture of carbon bi-sulphide and oil of turpentine.

The absolute alcohol gave the wire an increased resistance over air of .000064 ohm, corresponding to .19 of one per cent of the whole resistance. See Fig. 2.

The measurements for 90 per cent alcohol were more unsatisfactory than for most of the others on account of a much wider variation than usual between the individual measurements, both in case of the measurements in alcohol and in air. These measurements were made in two successive days from once filling the tube with alcohol. Four of the air measurements were made before the alcohol measurements and nine were made afterwards. The average of the eight measurements in alcohol was greater than that of the thirteen in air by .000024 ohm, but the irregularity in both series was so great that but little value should be attached to these results. The cause of this irregularity is unknown to me.

The curve for carbon bi-sulphide differed very little from that made for air. In this case, the air curve was made before filling the tube with the carbon bi-sulphide. Eleven measurements were made with this liquid on May 9th and 10th. The seven measurements made on the 9th all indicated a decreased resistance in the wire, though of a very small amount, while the four measurements on the 10th indicated an increased resistance of a somewhat greater amount. It seems probable that some contact resistance in the apparatus had increased by a small amount in the meantime.

The mixture of carbon bi-sulphide and oil of turpentine gave a decreased resistance to the wire of about .00003 ohm, or nearly .09 of one per cent.

With distilled water the irregularity was also very great. When the water was first poured in it seemed to give an increased resistance to the wire, but this increase seemed to disappear slowly as the water continued to stand in the tube. This seemed to me to be due to an increased conductivity of the water caused by the solution of salts of copper from the wire and tube.

The following tables give the figures relating to the different measurements as preserved in my note book. It will be observed that these measurements were made at all times of the day, so that any difference of temperature between the wire and the resistances used for comparison would not always affect the result in the same direction. The same resistance box and thermometer were used throughout.

DATE	HOUR	DIELECTRIC	TEMP.	RESIST- ANCE IN OHMS
1892 Feb. 22		Air	23.0	.03385
"	12 M.	"	22.8	.03380
"	2 P. M.	"	25.0	.03410
"	5 "	"	25.2	.03410
Feb. 23	9 A. M.	"	19.1	·03340
"	4 P. M.	"	21.9	.03370
"	6 "	"	23.0	.03385
Feb. 24	8.30 A. M.	"	17.7	.03320

DATE	HOUR	DIELECTRIC	TEMP.	RESIST-ANCE IN OHMS
Feb. 24	10 A. M.	Air	18.3	.03330
"	11.30 A.M.	"	20.0	.03350
"	3.15 P.M.	"	22.0	.03370
"	5 "	"	22.8	.03383
Feb. 25	10 A. M.	"	. 18.0	.03327
Feb. 26	9.30 "	"	19.5	.03347
"	3.20 P. M.	"	21.6	.03368
"	3.45 "	Wood Alcohol	21.1	.03366
"	5.20 "	"	21.7	.03367
"	9.00 "	"	22.6	.03377
Feb. 27	4.00 "	"	21.0	.03365
Feb. 28	9.30 A. M.	"	18.0	.03328
Feb. 29	10 "	"	19.6	.03343
"	2.30 P. M.	"	21.7	.03370
"	6.00 "	"	21.0	.03360
March 1	4.00 "	"	22.0	.03374
March 3	10.30 A. M.	"	19.9	.03350

Poured out wood alcohol and filled with petroleum.
Some wood alcohol left in the tube.

DATE	HOUR	DIELECTRIC	TEMP.	RESIST- ANCE IN OHMS
1892 March 4		Petroleum and Wood Alcohol	21.0	.03370
"		"	22.0	.03385
"	4.00 P. M.	"	22.6	.03388
"	5.00 "	"	23.0	.03392
March 5	9.00 A. M.	"	17.7	.03330
"	9.30 "	Air	19.0	.03340
"	9.45 "	Petroleum and Wood Alcohol	19.1	.03347
"	10.30 "	Air	21.0	.03365
"	10.45 "	Petroleum and Wood Alcohol	20.0	.03360

Disconnected apparatus and cleaned tube by draining
out the liquid and forcing air through it with a bellows.

DATE	HOUR	DIELECTRIC	TEMP.	RESIST- ANCE IN OHMS
March 6	7.45 P. M.	Air	24.1	.03418
" 7	8.15 A. M.	"	21.0	.03390
" "	9.30 "	"	22.0	.03400

Date	Hour	Dielectric	Temp.	Resist-ance in Ohms
March 7	10.45 A. M.	Air	23.1	.03418
" "	11.10 "	"	23.4	.03420
" "	3.00 P. M.	"	24.7	.03431
" "	3.50 "	"	25.0	.03434
" "	4.50 "	"	26.0	.03442
" 8	9.15 A. M.	"	19.9	.03374
" "	9.50 "	"	20.3	.03383
" "	10.30 "	"	20.8	.03384
" "	11.45 "	Petroleum*	22.2	.03390
" "	12.15 P. M.	"	22.0	.03390
" "	1.30 "	"	22.7	.03400
" "	2.10 "	"	23.0	.03406
" "	4.00 "	"	24.8	.03426
" 9	9.00 A. M.	"	20.0	.03370
" "	10.00 "	"	21.2	.03386
" "	11.15 "	"	21.7	.03392
" "	12.00 M.	"	22.0	.03396
" "	3.00 P. M.	"	23.1	.03408

* Whittier, Fuller & Co's. 150 ° Fire Test " Star " Kerosene.

Date	Hour	Dielectric	Temp.	Resistance in Ohms
March 9	5.00 P. M.	Petroleum	24.0	.03415
" 10	10.00 A. M.	Air	21.6	.03392
". "	11.00 "	"	21.7	.03394
" "	1.15 P. M.	"	21.8	.03395
" "	4.00 "	"	20.1	.03374
" 11	10.00 A. M.	"	20.6	.03382
" 12	2.10 P. M.	"	19.1	.03366
" 13	10.00 A. M.	"	17.8	.03350
" "	10.40 "	"	18.9	.03360
" 14	10.00 "	"	20.3	.03380
" 28	10.45 "	"	16.3	.03330
" "	11.20 "	"	16.3	.03330
" "	4.00 P. M.	"	17.0	.03340
" 29	10.30 A. M.	"	17.7	.03348
" 30	9.30 "	"	15.6	.03325
April 2	2.00 P. M.	"	17.9	.03349
" "	2.10 "	Petroleum	18.9	.03355
" "	4.40 "	"	17.7	.03343

DATE	HOUR	DIELECTRIC	TEMP.	RESISTANCE IN OHMS
April 2	5.30 P. M.	Petroleum	17.6	.03340
" 3	9.50 A. M.	"	13.6	.03293
" "	10.10 "	"	14.0	.03298
" "	10.50 "	"	15.0	.03310
" "	12.10 P. M.	"	16.6	.03330
" "	5.40 "	"	16.6	.03330
" 4	9.50 A. M.	Air	15.1	.03320
" "	11.40 "	"	15.1	.03315
" "	4.00 P. M.	"	18.9	.03364
" "	4·10 "	"	18.9	.03365
" "	4.45 "	"	19.2	.03367
" "	4.50 "	"	19.2	.03367
" "	5.30 "	"	19.3	.03368
" "	7.20 "	Petroleum	19.2	.03364
" 5	8.30 A. M.	"	13.6	.03298
" "	10.10 "	"	14.0	.03304
" "	11.00 "	"	14.6	.03310

The curves representing the above measurements for air and petroleum are given in Fig. 1.

The petroleum was mostly poured off and the tube filled with Wood Alcohol at 11.35 A. M.

DATE	HOUR	DIELECTRIC	TEMP.	RESISTANCE IN OHMS
April 5	12.00 M.	Wood Alcohol and Petroleum	14.8	.03308
" "	12.30 P. M.	"	15.3	.03320
" "	1.45 "	"	16.4	.03334
" "	3.00 "	"	17.6	.03352
" "	4.50 "	"	18.4	.03363
" "	5.30 "	"	18.9	.03368
" "	6.00 "	"	19.1	.03368
" "	7.35 "	"	19.0	.03364
" 6	8.30 A. M.	"	16.1	.03333
" "	9.45 "	"	17.1	.03345
" "	10.45 "	"	18.3	.03363
" "	11.45 "	"	18.7	.03370
" "	12.35 P. M.	"	19.0	.03372
" "	1.45 "	"	19.5	.03376
" "	3.30 "	"	20.7	.03395

2

DATE	HOUR	DIELECTRIC	TEMP.	RESIST-ANCE IN OHMS
April 6	4.40 P. M.	Wood Alcohol and Petroleum	21.5	.03405
" "	5.30 "	Air	21.9	.03395
" "	5.50 "	"	21.6	.03392
"	7.40 "	"	21.0	.03386
" "	9.20 "	"	15.3	.03318
" "	10.00 "	"	18.6	.03355
" 7	7.25 A. M.	"	15.2	.03316
" "	9.15 "	"	18.4	.03355
" "	10.05 "	"	19.0	.03362
" "	10.20 "	Wood Alcohol and Petroleum	18.0	.03357
" "	10.50 "	"	18.9	.03366
" "	11.30 "	"	19.1	.03368
" "	12.30 P. M.	"	19.6	.03373
" "	1.45 "	"	20.0	.03381
" "	4.00 "	Air	23.0	.03414
" "	5.30 "	"	22.1	.03405
" "	5.45 "	"	24.5	.03434
" "	8.00 "	"	21.1	.03390

It will be noticed that the first measurement after pouring the wood alcohol into the tube which had just contained the petroleum gave the same resistance that the wire had shown in petroleum, and that not until the fourth measurement, made over three hours after the wood alcohol was poured in, did the true resistance in the mixed liquid seem to be reached. This phenomenon was frequently observed by pouring off a liquid dielectric and making a measurement at once, and in nearly every case the resistance seemed to be either that given by the liquid dielectric at that temperature, or one between that of the liquid and the air resistance at the same temperature. From this, I was led to think that the thin layer of this dielectric in contact with the wire had a marked influence upon the wire's conductivity.* For this reason when the liquid was poured out the tube was usually carefully drained out and dried by blowing air through it with a bellows or drawing it through with a filter pump.

After the last measurement in air given above, the tube was filled with ordinary 90 per cent alcohol, and the following measurements were made.

---

* Since the above was in type, my attention has been called to the remarks made by Prof. von Helmholtz before the recent meeting of the British Association at Edinburgh, as reported in "The Electrician" of Aug. 12, 1892.

The substance of the report, so far as it applies to the question in hand, is that a mercury column in a glass tube may have a greater resistance due to the film of air adhering to the inside of the tube, even when the thickness of this film does not exceed .0005 of a wave length of light, and that the lowest resistance of the column may be obtained by allowing a drop of petroleum to spread itself over the inside of the tube. It has seemed to me possible that this phenomenon may be related to the one described above.

DATE		HOUR	DIELECTRIC	TEMP.	RESIST-ANCE IN OHMS
April 7		9.20 P. M.	Alcohol	19.0	.03365
"	8	7.30 A. M.	"	14.0	.03304
"	"	8.40 "	"	15.1	.03325
"	8	9.40 "	"	16.6	.03341
"	"	10.50 "	"	17.9	.03355
"	"	12.30 P. M.	"	19.0	.03370
"	"	1.35 "	"	19.9	.03381
"	"	3.00 "	"	21.0	.03395
"	"	4.05 "	Air	22.2	.03404
"	"	5.30 "	"	22.2	.03408
"	"	8.00 "	"	21.0	.03394
"	"	10.00 "	"	20.0	.03376
"	9	7.30 A. M.	"	15.6	.03328
"	"	8.40 "	"	16.3	.03339
"	"	9.10 "	"	17.6	.03351
"	"	9.30 "	"	17.9	.03352
"	"	10.40 "	"	18.6	.03360

Following these measurements were some made with gaseous dielectrics which will be described later. On April 20th, 21st, 22d, 24th, and 25th, twenty-six measurements were made in air, giving the curve shown in Fig. 2. These measurements are given in the table comparing air and illuminating gas. Directly after finishing the air and gas measurements the following measurements were made in absolute alcohol, and were compared with the air curve just mentioned.

DATE	HOUR	DIELECTRIC	TEMP.	RESIST- ANCE IN OHMS
April 27	2.50 P. M.	Absolute Alcohol	18.4	.03372
" "	3.35 "	"	18.8	.03377
" "	4.25 "	"	19.3	.03385
" "	6.00 "	"	19.9	.03390
" "	7.50 "	"	19.3	.03385
" 28	7.30 A. M.	"	11.9	.03300
" "	8.20 "	"	13.9	.03321
" "	9.10 "	"	15.9	.03343
" "	9.50 "	"	16.9	.03356
" "	10.30 "	"	17.5	.03364
" "	11.30 "	"	18.0	.03368
" "	12.10 P. M.	"	18.3	.03373
" "	1.35 "	"	18.8	.03377

Date	Hour	Dielectric	Temp.	Resistance in Ohms
April 28	2.40 P. M.	Absolute Alcoh'l	19.2	.03383
" "	3.20 "	"	19.9	.03392
" "	6.10 "	"	21.0	.03403
" "	7.45 "	"	20.6	.03400
" "	7.30 A. M.	"	16.9	.03355
" 29	8.50 "	"	19.0	.03380

After pouring out the alcohol, the tube was left filled with the vapor of alcohol and the following measurements were made, showing that the resistance returned to practically the former resistance in air.

Date	Hour	Dielectric	Temp.	Resistance in Ohms
April 29	9.50 A. M.	Air and Alcohol Vapor	20.0	.03388
"	10.37 "	"	20.9	.03398
"	11.30 "	"	21.9	.03410
"	12.15 P. M.	"	22.2	.03414
"	1.45 "	"	22.9	.03423
"	4.00 "	"	22.0	.03412
"	6.00 "	"	20.6	.03396
"	8.08 "	"	18.5	·03369

On May 6th, a new series of air measurements was begun. The contacts between the bridge and tube had been broken off and re-soldered in the meantime, so that no comparison can be made between these measurements and the preceding.

DATE	HOUR	DIELECTRIC	TEMP.	RESIST- ANCE IN OHMS
May 6	2.35 P. M.	Air	20.1	.03380
"	3.55 "	"	19.4	.03373
"	4.40 "	"	19.9	.03378
"	5.50 "	"	20.4	.03385
" 7	7.30 A. M.	"	15.9	.03329
"	9.00 "	"	17.9	.03353
"	4.00 P. M.	"	21.8	.03399
"	4.40 "	"	22.7	.03412
"	5.30 "	"	23.3	.03420
"	7.10 "	"	22.8	.03414
" 8	8.00 A. M.	"	13.4	.03300
"	8.55 "	"	14.9	.03317

Poured in about a spoonful of carbon bi-sulphide, and allowed to stand until the vapor should fill the tube.

DATE	HOUR	DIELECTRIC	TEMP.	RESIST-ANCE IN OHMS
May 8	10.55 A. M.	Air & Carbon Bi-sulphide Vapor	17.2	.03346
" 9	7.30 "	"	17.7	.03352
" "	9.20 "	"	19.8	.03377

As the vapor in the tube seemed to cause no sensible
deviation from the measurements in air, the tube was
filled with ordinary commercial carbon bi-sulphide and
the following measurements were made. It will be observed
that the measurements made on the 9th all indicate a
decreased resistance on account of the liquid, while those
made on the 10th indicate an increase. As some measurements made in air on the 10th also indicate an increase,
I have assumed that the action of the carbon bi-sulphide
upon the copper wire had changed its resistance by about
.00003 ohm, which seemed to be about the amount of the
permanent change. If this be true, the resistance in carbon
bi-sulphide is slightly less than in air. I hope, however, to
repeat these measurements during the coming year.

DATE	HOUR	DIELECTRIC	TEMP.	RESIST-ANCE IN OHMS
May 9	10.25 A. M.	Carbon Bi-Sulphide.	19.7	.03373
" "	11.30 "	"	19.8	.03375
" "	12.30 P. M.	"	20.0	.03378

DATE	HOUR	DIELECTRIC	TEMP.	RESISTANCE IN OHMS
May 9	1.55 P. M.	Carbon Bi-Sulphide	20.6	.03386
" "	2.45 "	"	21.1	.03392
" "	4.45 "	"	22.0	.03402
" "	6.00 "	"	21.5	.03397
" 10	7.30 A. M.	"	17.3	.03350
" "	8.30 "	"	19.0	.03370
" "	9.00 "	"	19.8	.03378
" "	9.35 "	"	20.3	.03385

May 11th to 17th, the following measurements were made with air, wood alcohol, benzine, and wood alcohol and benzine mixed.

DATE	HOUR	DIELECTRIC	TEMP.	RESISTANCE IN OHMS
May 11	8.55 A. M.	Air	15.6	.03330
" "	9.50 "	"	16.6	.03343
" "	11.05 "	"	17.2	.03352
" "	12.30 P.M.	"	19.2	.03375

Date	Hour	Dielectric	Temp.	Resistance in Ohms
May 11	2.15 p. m.	Air	20.5	.03391
" "	4.15 "	"	22.0	.03408
" "	5.35 "	Wood Alcohol	21.9	.03407
" "	7.40 "	"	20.4	.03390
" 12	7.25 a. m.	"	13.4	.03305
" "	8.20 "	"	15.0	.03323
" "	9.15 "	"	16.0	.03337
" "	9.55 "	Air	17.2	.03352
" "	10.20 "	Benzine	17.3	.03354
" "	11.30 "	"	18.6	.03367
" "	2.15 p. m.	"	19.0	.03372
" 14	7.30 a. m.	"	16.0	.03337
" "	8.30 "	"	16.9	.03348
" "	12.35 p. m.	"	18.7	.03370
" "	2.00 "	"	19.3	.03378
" "	5.40 "	"	20.1	.03390
" 15	9.05 a. m.	"	15.9	.03337

Poured out about half of the benzine and replaced by wood alcohol.

DATE	HOUR	DIELECTRIC	TEMP.	RESISTANCE IN OHMS
May 15	2.00 P. M.	Benzine and Wood Alcohol	20.2	.03390
" "	7.30 "	"	20.2	.03389
" 16	8.15 A. M.	"	16.0	.03342
" "	9.05 "	"	16.9	.03352
" "	10.00 "	"	17.9	.03363
" "	12.00 M	"	19.0	.03377
" "	1.45 P. M.	"	19.8	.03385
" "	3.30 "	"	20.9	.03397
" "	5.30 "	"	21.8	.03410
" 17	9.50 A. M.	Air	21.7	.03405
" "	12.20 P. M.	"	22.9	.03413
" "	1.35 "	"	23.8	.03430

In the above measurements the increase of resistance caused by mixing the two dielectrics is very noticeable.

The measurements with distilled water, as before stated, were too irregular to enable one to decide upon the position in the series of this dielectric. They seem to show, however, that until its conductivity has been increased, it causes an increased resistance in the wire.

Filled with distilled water at 1.45 P. M., May 17.

DATE	HOUR	DIELECTRIC	TEMP.	RESISTANCE IN OHMS
May 17	3.05 P. M.	Dist. Water	24.4	.03444
" "	3.55 "	"	25.0	.03445
" "	5.00 "	"	26.0	.03445
" "	6·00 "	Air	27.1	.03475
" "	8.00 "	"	25.1	.03450
" 18	8.15 A. M.	"	21.0	.03403
" "	9.25 "	"	22.4	.03418
" "	10.06 "	"	23.0	.03427
" "	10.15 "	Dist. Water	21.1	.03412
" "	11.45 "	"	23.1	.03433
" 19	9.45 "	Air	22.0	.03415

Filled tube with mixture of carbon bi-sulphide and oil of turpentine.

DATE	HOUR	DIELECTRIC	TEMP.	RESISTANCE IN OHMS
May 19	10.45 A. M.	Carbon Bi-sulphide and Turpentine Oil	23.8	.03432
"	11.30 "	"	24.9	.03445

DATE	HOUR	DIELECTRIC	TEMP.	RESISTANCE IN OHMS
May 19	12.15 p. m.	Carbon Bi-sulphide and Turpentine Oil	25.7	.03454
"	1.45 "	"	26.9	.03467
"	2.35 "	"	28.0	.03485
"	3.35 "	"	29.2	.03500

Regarding the conductivity of the wire in air as unity, the conductivity in the liquids examined was as follows :

Petroleum - - - - - - -	1.0018
Carbon Bi-Sulphide and Oil of Turpentine -	1.0009
Carbon Bi-Sulphide, uncertain, apparently -	1.+
Wood Alcohol - - - - - - -	.9998
Benzine - - - - - - - -	.9994
Wood Alcohol and Benzine - - - -	.9985
Absolute Alcohol - - - - - -	.9981
Wood Alcohol and Petroleum - - - -	.9973
Distilled Water, uncertain, apparently - -	1.—

I have been unable in the limited time at my command to find any fixed relation between the conductivities of the wire in the various dielectrics and any other known properties of the dielectrics. The tables for specific inductive capacity seem to me, on account of the very contradictory results of different observers, to be of little value, and the theoretical relation which it seems should exist

between the specific inductive capacity and the coefficient
of refraction of light still seems to be contradicted oftener
than it is verified by experiment. In the hope that I
might find some relation between the refractive indices of
the liquids used and their dielectric properties as shown
in these experiments, I had the refractive indices of the
same liquids determined in this laboratory by Mr. Murphy, a very careful and conscientious observer. Taking
the refraction of air as unity, Mr. Murphy obtained the
following results for the D line :

Petroleum, poured off the mixture of wood alcohol
  and petroleum - - - - - - 1.440

Petroleum, pure - - - - - - - 1.435

Benzine, from the mixture of wood alcohol and
  benzine - - - - - - - 1.405

Benzine, pure - - - - - - - 1.403

Absolute Alcohol - - - - - - 1.363

Wood Alcohol - - - - - - - 1.344

Wood Alcohol, from which petroleum had been
  poured off - - - - - - - 1.340

It will be seen at once that the order of arrangement
of the liquids is very different in the two tables. While
petroleum has the highest refractive index of any single
liquid used, and has likewise the highest place in regard
to the dielectric property under consideration, absolute
alcohol, which is at the other extreme in its dielectric
behavior has a higher refractive index than the wood
alcohol, which differed but little from air in its dielectric
action upon the wire.

I hope to take up this question again next year with increased facilities for work, and to, at least, accumulate data in regard to a much larger number of dielectrics. There are several other important questions in this connection, the answers to which I have not yet had time to seek. Among them may be mentioned the question as to the influence of the surface condition of the wire used. In the experiments which I have thus far made, the surface of the copper wire was much oxidized. I hope to repeat some of the same experiments with a polished wire, and with wires covered with various insulators. The influence of the size of the wire also remains to be studied. Possibly, by this means one may be able to determine whether the phenomenon observed is due to some effect throughout the body of the dielectric, or is confined to the surface of the conductor. The effect of alternating currents through different dielectrics will also furnish an interesting field for investigation.

I may say in this connection that I have several times attempted to observe similar phenomena with wires coiled in various ways, and measured in air and in other dielectrics, but so far, I have found a definite change in resistance only when the current was sent through the tube and wire as previously described. In using a coil of wire instead of the tube for the outer conductor, however, the resistance is so much increased that the same change in resistance would not be noticeable with the apparatus at my command.

EFFECTS OF SOME GASEOUS DIELECTRICS UPON CONDUCTIVITY

Early in the course of the experiments which have been described, it was observed that after pouring out the liquid dielectric some time was required for the resistance of the wire to return to its normal condition in air. This suggested the thought that the vapor of the liquid left in the tube might be the cause of this phenomenon. The following experiments were accordingly made with gaseous dielectrics.

The first gaseous dielectric definitely tested was the burning gas used in the laboratory. This gas is made from gasoline by a machine on the grounds, and consists of the vapor of gasoline mixed with air. Its sp. g. is somewhat greater than air. It was allowed to enter the tube direct from the pipes in the room, and after it had driven the air out until the gas was coming through freely, the tube was closed and the measurements were made. The first series of measurements, made April 9th to 13th, showed that the resistance in the gas was greater than in air, but the measurements in both dielectrics were very irregular. Thinking that this must be due to irregularity in the contact resistances, the apparatus was taken apart, all plugs and wire contacts were carefully cleaned, and the apparatus was again connected for work. The result was the most consistent series of measurements made at any time.

DATE	HOUR	DIELECTRIC	TEMP.	RESIST- ANCE IN OHMS
April 20	8.15 A. M.	Air	16.3	.03340
" "	10.30 "	"	18.1	.03364
" "	11.30 "	"	18.1	.03363
" "	12.40 P. M.	"	18.0	.03362
" "	3.55 "	"	19.8	.03383
" "	5.40 "	"	19.4	.03378
" "	7.45 "	"	19.4	.03378
" "	9.00 "	"	19.1	.03374
" 21	8.05 A. M.	"	17.5	.03355
" "	8.55 "	"	18.8	.03372
" "	10.00 "	"	19.7	.03382
" "	11.30 "	"	21.1	.03400
" "	12.10 P. M.	"	21.9	.03410
" "	2.10 "	"	22.7	.03420
" "	3.10 "	"	22.8	.03420
" "	5.30 "	"	22.0	.03412
" 22	8.15 A. M.	"	17.7	.03357
" "	9.20 "	"	18.8	.03372

3

DATE	HOUR	DIELECTRIC	TEMP.	RESIST- ANCE IN OHMS
April 22	10.15 A. M.	Air	19.4	.03378
" "	10.30 "	Laboratory Gas	20.1	.03394
" "	11.15 "	"	19.8	.03390
" "	12.15 P. M.	"	19.7	.03386
" "	12.30 "	"	20.0	.03393
" "	1.30 "	"	21.0	.03403
" "	3.10 "	"	21.9	.03414
" "	5.10 "	"	22.9	.03427
" "	7.00 "	"	22.8	.03426
" 23	7.25 A. M.	"	12.0	.03295
" "	8.30 "	"	14.0	.03320
" "	9.00 "	"	15.0	.03332
" "	8.45 P. M.	"	18.4	.03372
" 24	7.40 A. M.	"	15.3	.03334
" "	8.10 "	"	16.1	.03346
" "	9.25 "	"	17.6	.03362
" "	11.00 "	"	18.7	.03377
" "	12.20 P. M.	"	18.8	.03377

The gas was blown out by air at 1.40 P. M. and the tube allowed to stand until 2.10, when air was again blown through it.

DATE	HOUR	DIELECTRIC	TEMP.	RESIST-ANCE IN OHMS
April 24	3.00 P. M.	Air	18.9	.03373
" "	5.40 "	"	19.2	.03377
" "	7.35 "	"	18.8	.03370
" 25	7.15 A. M.	"	11.4	.03283
" "	8.45 "	"	13.8	.03313
" "	9.20 "	"	14.4	.03320
" "	10.00 "	"	14.8	.03322

The curves representing the above measurements are shown in Fig. 3. Of the 17 measurements in gas, only one point is more than .00001 ohm from the curve, while the nearest distance of any gas point from the air curve is equivalent to a resistance of .00004 ohm, and the average distance of the gas points from the air curve is .000058 ohm. This would make the resistance of the wire in gas 1.0017 times its air resistance. Without moving the apparatus, a few spoonfuls of sulphuric ether were poured into the tube, the stop-cocks closed and the tube allowed to stand until it was filled with the vapor.

Date	Hour	Dielectric	Temp.	Resist-ance in Ohms
April 25	12.00 M.	Ether Vapor and Air	15.8	.03337
" "	1.50 P. M.	"	16.1	.03348
" "	3.10 "	"	17.0	.03358
" "	3.55 "	"	17.6	.03364
" "	5.00 "	"	18.2	.03373
" "	6.00 "	"	19.2	.03384
" "	7.25 "	"	19.0	.03382
" 26	7.30 A. M.	"	10.8	.03282
" "	8.25 "	"	12.8	.03307
" "	9.25 "	"	14.1	.03323
" "	10.30 "	"	15.0	.03334
" "	12.00 M.	"	15.4	.03337
" "	1.45 P. M.	"	16.2	.03352

The above measurements compared with the same air curve indicate an increase of resistance due to the ether vapor of .000083 ohm or .25 of one per cent. This is a greater variation than was given by any single liquid used. After the ether vapor had been drawn out of the tube by passing a current of air through it for more than an hour, the resistance fell but little from the above, showing the great effect due to a very small quantity of ether vapor. By washing the tube out with alcohol, the resistance of the wire fell to nearly its former value, as shown in the table for measurements in absolute alcohol made on April 27th, 28th, 29th and 30th, and which were compared with the same air curve used for Fig. 3.

The effect of chloroform vapor is shown by the following table :

DATE	HOUR	DIELECTRIC	TEMP.	RESISTANCE IN OHMS
May 4	4.00 P. M.	Air	22.1	.03428
" "	5.55 "	"	21.2	.03418
" "	7.00 "	"	20.0	.03404
" "	9.00 "	"	18.9	.03393
" 5	7.30 A. M.	"	14.9	.03346
" "	8.30 "	"	16.8	.03370
" "	9.13 "	"	17.9	.03382
" "	9.35 "	"	18.4	.03387
" "	9.50 "	Chloroform Vapor and Air	19.0	.03397
" "	10.20 "	"	19.2	.03400
" "	11.05 "	"	20.0	.03412
" "	11.50 "	"	20.4	.03416
" "	12.30 P. M.	"	21.1	.03423

DATE	HOUR	DIELECTRIC	TEMP.	RESISTANCE IN OHMS
May 5	2.30 P. M.	Chloroform Vapor and Air	20.9	.03420
" "	3.40 "	"	20.2	.03413
" "	7.20 "	"	18.4	.03393

In the above measurements the tube was not moved and no contact was disturbed in any way. Only a few teaspoonfuls of the chloroform were poured into the tube and the first measurement was made ten minutes afterward. The eight measurements made in each dielectric are thoroughly consistent and indicate an increased resistance in the chloroform vapor of nearly .17 of one per cent.

A few measurements made in the vapor of carbon bisulphide on May 9th showed no appreciable difference from the air resistance. On May 10th an attempt was made to find the effect of rarefied air as a dielectric, but the apparatus was found to leak air so badly that no very near approach to a vacuum could be reached The measurements marked " Vacuo " in the table were made while the tube was nearly hermetically sealed and a large sized Chapman filter pump was exhausting the air. The water pressure upon the pump was very heavy, and it would exhaust a glass tube so that the mercury would rise in it to within one cm. of the barometric height. It is accordingly probable that there were only a few cm. of air in the tube. It is probable, however, that the pressure was not the same for the different measurements.

Date	Hour	Dielectric	Temp.	Resistance in Ohms
May 10	11.10 A. M.	Vacuo	21.1	.03398
" "	11.50	"	21.5	.03402
" "	1.20 P. M.	"	22.0	.03412
" "	1.35 "	Air	22.2	.03408
" "	2.10 "	"	22.7	.03415
" "	3.10 "	Vacuo	22.4	.03414
" "	3.25 "	Air	22.9	.03418
" "	4.15 "	Vacuo	22.9	.03420
" "	4.55 "	Air	23.4	.03425
" "	5.20 "	Vacuo	23.7	.03428
" "	5.50 "	Air	23.9	.03428
" "	7.30 "	"	23.0	.03418
" 11	7.30 A. M.	"	13.4	.03304
" "	8.10 "	Vacuo	14.5	.03315
" "	8.55 "	Air	15.6	.03330
" "	9.50 "	"	16.6	.03343
" "	11.05 "	"	17.2	.03352
" "	12.30 P. M.	"	19.2	.03375

DATE	HOUR	DIELECTRIC	TEMP.	RESIST-ANCE IN OHMS
May 11	2.15 P. M.	Air	20.5	.03391
" "	4.15 "	"	22.0	.03408

The air measurements all lie very close to the curve representing the average, while of the measurements in rarefied air, all but one indicate an increased resistance, but are too irregular to be of any quantitative value.

Taking the conductivity of the wire in air as unity, the conductivities in the other gases tested are approximately as follows :—

Alcohol Vapor -   -   -   -   -   -   - .99949
Chloroform Vapor   -   -   -   -   - .99830
Gasoline Burning Gas   -   -   -   - .99820
Sulphuric Ether Vapor -   -   -   -   - .99750
Carbon Bi-Sulphide Vapor, approximately, -   1.
Rarefied Air, less than   -   -   -   -   - 1.

I am not aware that any variation in the dielectric properties of gases corresponding to the above results have ever been observed. I had no means of measuring the refractive indices of the vapors used, and of thus computing their theoretical specific inductive capacity. Taking the refractive indices as given by Lorenz, Landolt and Börnstein's Physikalisch-Chemische Tabellen, Berlin, 1883, the theoretical specific inductive capacities of the vapors used are, in terms of air :

Alcohol Vapor - - - - - -	1.00120
Chloroform Vapor - - - - - -	1.00234
Carbon Bi-Sulphide Vapor - - - -	1.00242
Ether Vapor - - - - - - -	1.00321

In this table, the alcohol, chloroform and ether vapors are arranged in the same order as they would be with reference to this dielectric property, but the carbon bi-sulphide vapor is not. Probably, no value should be attached to this comparison, for the reason that the vapors were mixed with air in these determinations, and their density was, perhaps, very different from that at which their refractive indices were determined.

I hope to pursue these investigations much farther during the coming year, and to accumulate data from which it may be legitimate to make comparisons.

JUNE 8, 1892.